Where The River Meets The Sea

Jacques Levy, '55

Where The River Meets The Sea

RiverSea Press

Coastal Landscapes
1980-1998

Jacques Levy

Where The River Meets The Sea
Landscapes
1980-1998
Jacques Levy

First Edition

Published by:
RiverSea Press
P.O. Box 105
Occidental, CA 95465   USA

Copyright © 1998 by Jacques Levy

All rights reserved.

No part of this book may be reproduced in any form without the written permission of the publisher, except for brief quotes in a published review.

Library of Congress Catalogue Card Number: 98-96139

ISBN 0-9663863-0-2

Printed in USA   First Printing

10 9 8 7 6 5 4 3 2

Front Cover: *December 26, 1997* Watercolor 12x16 in.

Frontispiece: *#1/1980* Oil on Panel 30x36 in.

Logo: *March 20, 1995* Oil on Panel 32x40 in.

Contents

7 Dedication

8 Acknowledgements

9 Introduction by the Artist

12 Notes on the Plates

13 Oil Paintings

33 Watercolors

53 Pen and Ink Drawings

For Dotty

Where the skyslopes fall

And the river meets the sea

There we ever were

And so shall always be

Acknowledgements

This book owes much of its development to the many friends and acquaintances who generously and enthusiastically offered their time and creative input.

First and foremost, I want to thank my partner, Dotty Joos, for critical insights that improved the text and layout and for her subtle and wily humor which peppers as it illuminates.

Narrowing the choice of works was a major challenge, so I called upon people with experience in the visual arts. My son, filmmaker Abe Levy provided valuable formal reviews of potential selections; old friends and fellow artists Byron Randall and Pele deLappe examined the mock-up and encouraged the project; painter and neighbor Charles Beck offered sensitive ideas on overall design and selections; and artist Mike Kosturos helped convince me that creating this monograph would be a natural resolution of my wish to share my work with others.

My deep appreciation goes out to those who provided information, criticism and support: Francine Allen, Lorraine Almeida, Barbara Baer, Wilder Bentley, Jim Colvin, Amie Hill, Lionel Gambill, Janet Greene, Lisa Hammon, Daedalus Howell, Doret Kollerer, my daughter Aerielle Levy, Jane Love, Anita Peery, Sienna S'Zell, Salli Rasberry, Dolores Richards and David Thatcher.

Dietmar Krueger's photographs of the paintings were done after his numerous tests and exceptional efforts overcame unexpected problems of specularity. Lance Bollens and Bruce McLeester of Summerfield Graphics patiently guided me through the complexities of color separations, and Lorraine Fiamengo brought the text files into conformity with required protocols. Gretchen Singer and Paramesh Adhikari of Singer Printing raised my confidence in the eventual quality of this book with detailed discussions and examples of their work. Finally, a word of thanks to others who helped along the way: photographers Richard Allen and Scott McCue, and printers Pat Barlow of Barlow Printing, Steve Coburn of Chromagraphics and Bob Lorey of Lithocraft.

Introduction by the Artist

Or‑gan´ic, a. [L. *organicus*, from *organum*, implement.]
    2. inherent; inborn; constitutional.
    3. organized; systematically arranged.

This essay gives me the opportunity to describe some of the more important influences which combined to help bring the body of work shown in this book to its present state and thereby, hopefully, to guide the reader to a better understanding of it.

Being a painting student at the San Francisco Art Institute in the late '60s meant being very much influenced by The New York School. In fact, my teachers were mostly second-generation abstract expressionists, as exponents of the School came to be known. Some of them had been initiated into the mysteries of that way of painting by Bay Area greats Mark Rothko and Clyfford Still.

At first I didn't have a clue about the significance of abstract expressionism, then not much more than a clue for some time afterwards. At the visual level, it seemed like chaos taken to the limit, oblivion elevated to a sacrament. "Organic" and "existential," though, were the verbal descriptions of choice at the time. Nevertheless, at the intersection of process and feeling, it was obvious that abstract expressionism pointed towards some kind of freedom for the painter: freedom to explore and express the limitless possibilities of a plastic medium—and to get lost in it.

As I struggled for clarity and meaning during my second post-graduate year, a San Francisco Sufi had an unexpected impact on the direction of my work when he introduced me to the notion of *lights and darks in balance.* At about that time, I was also exposed to the thinking of a New York art critic who had said that some of Jackson Pollack's paintings exhibited what he characterized as "classic chiaroscuro." Thus I slowly awakened to a major issue in the theory and practice of my chosen craft.

Where The River Meets The Sea

When my student work culminated in a disappointing solo show in San Francisco in 1970, I was destined to spend the next ten years wandering in the spectral domain of vision and hallucination. But in 1980, an unexpected painting (see frontispiece) emerged which somehow forecast years of work to come. I thought then that a new path had opened to me: to explore how the medium with infinite possibilities could lead to the creation of paintings with the visual logic of the physical world while being developed entirely in the studio. My landscapes were born.

The oil paintings, watercolors and drawings in this book are shaped largely by memory, imagination and the visual cues provided by the paint itself. Commencing with work begun around 1994, they were also informed by a color triangle developed by eminent American theorist Faber Birren (*Principles of Color,* 1987). Based on a similar triangle invented by German physiologist Ewald Hering (ca. 1878), Birren's diagram, with any pure hue at one angle and black and white at the other two, demonstrates how all potential color modifications (tints, tones, and shades) can develop within its borders. Awareness of these modified colors contributed unanticipated new dimensions to my work because of their power to express the range of diurnal and seasonal changes in the natural world as well as the variability of human experience.

Already interested in the issue of lights and darks, I was excited by Birren's triangle because it provided me with a theoretical roadmap for the integration of color with black and white. When I began to use it systematically and combined it with elements of theory advanced in M.E. Chevreul's *Principles of Harmony and Contrast of Colors* (1839) and in Albert Munsell's *Grammar of Color* (1921), I was taken full circle to the words of the Sufi and to a growing conviction that I had found a reliable and satisfying principle by which to organize my work.

While I can explain a method for dealing with color, words cannot account sufficiently for the evolution of space and form in paintings in which process plays a decisive role. So much depends on chance, intuition and the mysteries of expressive vision. I'm sometimes asked, "What will the painting depict?" I usually reply that I won't know until it is done. But this question may be partly resolved by considering the visually paradoxical equivalence of paint with space and form, a principle clarified during my visit to London's Tate Gallery during the spring of 1996.

Introduction by the Artist

I had traveled to London to see the legendary painting collection of J. M. W. Turner (1775-1851), but even years of exposure to monographs of his work had hardly prepared me for the scope and power of his expressive vision. After studying Turner's oils for several hours, I was finally exhausted by my own excitement and so decided to take a break.

In an adjoining gallery within the Tate, I soon found myself absorbed by William McNeal Whistler's (1834-1903) blue night view of *Old Battersea Bridge* (1872-75). Along came a tour group with a bespectacled docent at its head. He and the group decided to study the Whistler and I decided to hear what the docent would say about it. He explained that the scene was not understood by the English of the 19th century because they were accustomed to the mimetic conventions of painterly representation. In other words, the English did not think the painting *looked* like the bridge. I was puzzled by the docent's explanation because it never occurred to me that it was anything but the bridge. And so I learned how much art invention has done for human perception through time, or at least for *my* perception.

Back at the Turner Collection later on, I was struck by how much less mimetic Turner's late paintings were than his early ones, and that reminded me of once having read a Turner exhibition catalogue, *Imagination and Reality*, in which the writer said that Turner's late work had prefigured abstract expressionism! Something like that had also been said about Albert Ryder.

And so I see that when I paint, I partake of the same mystery that joins Turner and Ryder to the abstract expressionists. I simply spread pure paste pigment until the texture, color, and viscosity of the medium and my increasing visual engagement lead to a critical intersection: the moment that full emotional commitment transforms effort into the certainty of the creative experience; the unexpected, thrilling and astonishing moment which coincides with the transformation of the painting from a work-in-progress into a visual space embraced by the mind and beyond time.

Jacques Levy
Occidental, California
March, 1998

Notes on the Plates

<u>1980-1983</u>

Paintings in this group (frontispiece plus plates 2-4) were created in Tomales, California. Scumbling layer upon layer of pigment with a palette knife, I created textures which sometimes mimicked natural forms. My color palette was haphazard but tended generally toward high chroma reds and yellows combined with low-value earth tones and black. In retrospect, I see that I favored pure hue and shades.

<u>1984-1989</u>

This period is not represented in the monograph because I had moved to Occidental and had gotten involved in peace and social justice issues. My work was confined to pen and ink drawings and black and white watercolors, a continuation of efforts begun years ago which concluded in a published collection of figurative imagery (see *Interior Exile, Works on Paper, 1966-1989*).

<u>1990-1991</u>

Plates 5-7 represent a period of very enthusiastic artistic renewal and a return to full-time work as a painter. Adding to textural discoveries made in previous years, I began to develop a fondness for stirring the colors around in the painting for days or weeks on end.

<u>1994-1998</u>

Plates 8-19, the painting on page 13, and all the watercolors were done after a move to a new residence and studio after the deaths of my father and eldest daughter in 1992. They mark the beginning of my interest in color theory, as well as my first intentional use of tints and tones.

Oil Paintings

Where The River Meets The Sea

1. Overleaf: *April 21, 1995* Oil on Panel 32x40 in.

2. *#1/1982* Oil on Board 20x24 in.

Oil Paintings

3. *#3/1982* Oil on Board 20x24 in.

Where The River Meets The Sea

4. *May 1983/#2* Oil on Board 20x26 in.

Oil Paintings

5. *March 22, 1991* Oil on Board 20x26 in. Collection Dorothea Joos

Where The River Meets The Sea

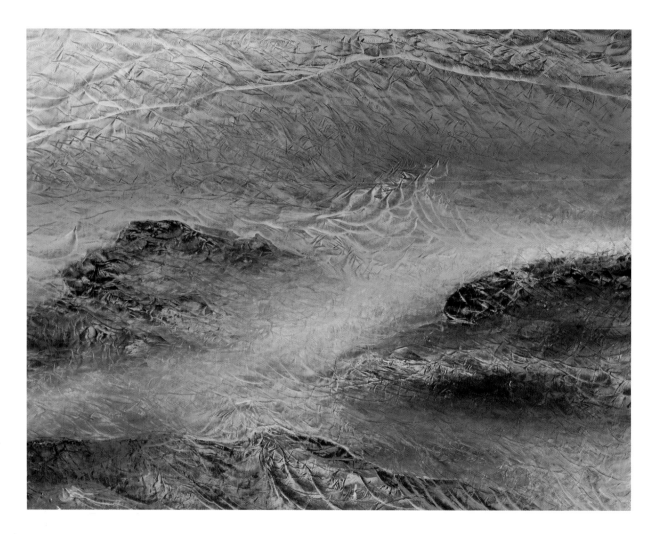

6. *March 30, 1991* Oil on Board 20x26 in.

Oil Paintings

7. *April 18, 1991* Oil on Board 20x26 in. Collection Avrom Lefkowitz

Where The River Meets The Sea

8. *October 25, 1994* Oil on Board 28x36 in.

Oil Paintings

9. *November 8, 1994* Oil on Board 28x36 in.

Where The River Meets The Sea

10. *November 27, 1994* Oil on Board 28x36 in.

Oil Paintings

11. *December 11, 1994* Oil on Board 28x36 in.

Where The River Meets The Sea

12. *February 24, 1995* Oil on Panel 32x40 in.

Oil Paintings

13. *March 12, 1995* Oil on Panel 32x40 in.

Where The River Meets The Sea

14. *March 20, 1995* Oil on Panel 32x40 in.

Oil Paintings

15. *April 8, 1995* Oil on Panel 32x40 in.

Where The River Meets The Sea

16. *December 15, 1996* Oil on Panel 32x40 in.

Oil Paintings

17. *April 16, 1997* Oil on Panel 32x40 in.

Where The River Meets The Sea

18. *November 30, 1997* Oil on Panel 32x40 in.

Oil Paintings

19. *March 4, 1998* Oil on Panel 32x40 in.

Where The River Meets The Sea

20. *January 27, 1998* Oil on Panel 32x40 in.

Where The River Meets The Sea

21. Overleaf: *November 20, 1997* Watercolor 12x16 in.

22. *November 17, 1994* Watercolor 15x20 in.

Watercolors

23. *November 22, 1994* Watercolor 15x20 in.

Where The River Meets The Sea

24. *December 10, 1994* Watercolor 15x20 in.

Watercolors

25. *December 22, 1994* Watercolor 15x20 in.

Where The River Meets The Sea

26. *May 26, 1995* Watercolor 15x20 in.

Watercolors

27. *June 19, 1995* Watercolor 15x20 in.

Where The River Meets The Sea

28. *June 18, 1996* Watercolor 14x20 in.

Watercolors

29. *June 26, 1996* Watercolor 14x20 in.

Where The River Meets The Sea

30. *November 17, 1996* Watercolor 10x14 in.

31. *November 22, 1996* Watercolor 10x14 in.

Where The River Meets The Sea

32. *September 30, 1997* Watercolor 12x16 in.

Watercolors

33. *November 4, 1997* Watercolor 12x16 in.

34. *December 25, 1997* Watercolor 12x16 in.

Watercolors

35. *December 26, 1997* Watercolor 12x16 in.

Where The River Meets The Sea

36. *January 3, 1998* Watercolor 12x16 in.

Watercolors

37. January 23, 1998 Watercolor 12x16 in.

Where The River Meets The Sea

38. *January 25, 1998* Watercolor 12x16 in.

Watercolors

39. *January 28, 1998* Watercolor 12x16 in.

Where The River Meets The Sea

40. *February 10, 1998* Watercolor 12x16 in.

Pen & Ink Drawings

41. Overleaf: *September 27, 1997* Pen and Ink 9x12 in.

42. *April 24, 1996* Pen and Ink 9x12 in.

43. *May 7, 1996* Pen and Ink 9x12 in.

44. *May 12, 1996* Pen and Ink 9x12 in.

45. *June 17, 1996* Pen and Ink 9x12 in.

Pen and Ink Drawings

46. *June 19, 1996* Pen and Ink 9x12 in.

47. *July 15, 1996* Pen and Ink 9x12 in.

48. *July 29, 1996* Pen and Ink 9x12 in.

49. *August 5, 1996* Pen and Ink 9x12 in.

Book Design: RiverSea Press, Occidental, CA
Photography: Dietmar Krueger, Bolinas, CA
Scans & Separations: Summerfield Graphics, Santa Rosa, CA
Pre-Flight Graphics: Fiamengo & Friends, Healdsburg, CA
Printing & Binding: Singer Printing, Petaluma, CA

Printed on Lustro Creme
Typeset in Palatino

Jacques Levy, April 1998